Guns on the Atlanti
1942-1945

With a mighty fireball, the shell leaves the barrel of a 34 cm railroad cannon. (BA)

Karl-Heinz and Michael Schmeelke

Schiffer Military/Aviation History
Atglen, PA

Bibliography and Acknowledgments

Federal Archive Files—Military Archives, Freiburg.
Delefosse, Atlantikwall-Heer, Abbeville, 1988.
Rolf, r., Der Atlantikwall, Amsterdam, 1983.
Zimmermann, Der Atlantikwall, Vol. II, 1989.
Leitfaden der Marineartillerie, Vol. II, Berlin, 1940.
Farmbacher, Lorient, Weissenburg, 1956.
Senger & Etterlin, Die deutschen Geschütze, Munich, 1960.
Chantal & Yannick Delefosse Archives, R. H. Zimmermann,

The drawings were made admirably by Chantal and Yannick Delefosse.

Photos

Federal Archive, Koblenz
Zimmermann
Author's archives: all other photos

Our special thanks go to the former members of the Navy: K. H. Riecken (MAA 240), Franz-Jos. Pötz (MAA 605), Friedrich Kuhr (MAA 264), Alfred Uher (MFAA 807)

COVER PICTURE:
A 12.5 cm Cannon Howitzer 433/1 (r) in an open gun bed.

A war correspondent films the process of aiming a 15.5 cm Field Cannon 418 (f). The aiming gunner, at far right, receives the shot values from the command post through headphones. The elevation gunner, at his left, cranks the barrel to the ordered height. The aiming optics can be seen in front of the aiming gunner.

Foreword

On the Atlantic Wall, the German fortifications that stretched along the coasts of Holland, Belgium and France as of 1942, a great number of different guns saw service.

Along with guns made in Germany, a much greater number of foreign guns, that had been captured by the Wehrmacht from 1940 on, were used.

In this volume, the guns most often used on the Atlantic Wall, from 5 cm to 34 cm calibers, will be described technically and in terms of their service.

Translated from the German by Ed Force

Printed in China.
ISBN: 0-7643-0572-7

This book was originally published under the title, *Waffen Arsenal-Geschütze am Atlantikwall 1942-1945* by Podzun-Pallas Verlag.

We are interested in hearing from authors with book ideas on related topics.

Published by Schiffer Publishing Ltd.
4880 Lower Valley Road
Atglen, PA 19310
Phone: (610) 593-1777
FAX: (610) 593-2002
E-mail: Schifferbk@aol.com
Please write for a free catalog.
This book may be purchased from the publisher.
Please include $3.95 postage.
Try your bookstore first.

History of the Atlantic Wall

In this volume we shall describe several guns, from 5 cm to 15.5 cm caliber, that were used in great numbers on the Atlantic Wall.

In addition, we shall show various heavy cannons up to 34 cm caliber that, if only individually, were used for coastal defense by the naval artillery and Army coast artillery units.

When Normandy was invaded on June 6, 1944, it was mainly the light and medium guns that opposed the Allied landing troops.

The basic concept for the construction of a coastal defense line was the avoidance of a two-front war, which could be foreseen after the entrance of the USA into the war on December 7, 1941.

For this reason, the Wehrmacht High Command (OKW) gave out the order one week later, on December 14, 1941, to build a "New West Wall".

This was supposed to stretch from Kirkenes in northern Norway to Hendaye on the Spanish border of France.

From fixed concrete positions, an attacker could then be fought off with as few German losses as possible.

To secure the Norwegian coast, the Wehrmacht transferred a great number of gun batteries there as early as 1940.

The Norwegian coastal batteries that existed were put to use by the German Army and Navy, sometimes after modernization. Here too, captured French and British weapons were used.

On the French Channel clast near Calais too, a great number of batteries had been stationed to support Operation "Sealion" (see also Weapon Arsenal S-22).

The other Atlantic coastal areas were secured by infantry divisions, but as of the summer of 1941 they were gradually to be withdrawn to serve on the Russian front, and were partially replaced by exhausted units recognized as needing recuperation. In fact, the "New West Wall" was erected by 1942 only in the areas named above, as well as in the occupied British Channel Islands and around the seaports used by the Navy, sych as Ijmuiden, Brest, Lorient, etc. In 1942 the Allies made the first moves against Western Europe, which was occupied by German troops. In the night of February 27-28, 1942, a British commando troop attacked the radar station at Bruneval, located on the steep coast in the vicinity of Le Havre. Here the British captured not only important components of the radar apparatus, but also two prisoners, who were immediately taken to Britain on a high-speed boat.

The Oerlandet Battery near Trondheim, Norway, was equipped with a three-gun 28-cm turret from the cruiser Gneisenau.

A 17 cm SKL/40 of the "Jade West" naval battery in an open position near Lorient.

One month later the U-boat harbor of St. Nazaire was attacked. Again the commando troop was able to overcome the German defense and land in the harbor, where extensive explosive and destructive tasks were undertaken.

On August 19 of the same year, 6000 Canadians of the 2nd Canadian Division landed on the beach at Dieppe, with strong air support.

The main target of this landing was the testing of new equipment, such as landing craft and tanks, as well as the capture of components of additional radar devices from the Pourville station.

This operation became a catastrophic failure. None of the goals could be achieved. The landing craft were able to reach the unfortified coast, but immediately came under the concentrated fire of the German coastal batteries.

Many of the new Churchill tanks could not climb the gravel beach typical of Dieppe and were stuck at the water line, making easy targets.

The capture of the radar station also failed. Immediately after the attack on the Bruneval station, all security measures at all radar stations were strengthened considerably. Thus the Canadians did not succeed in getting into the station, and with things going wrong, they had to withdraw to the beach again.

In all, the Canadians lost over 4000 men at Dieppe.

On the basis of this bitter experience, the Allies improved their preparations for an invasion of the continent immensely.

From then on, any beach that might be invaded by commando troops was inspected nightly, measured precisely, and subjected to ground testing.

As a result of these attacks, Hitler ordered on September 29, 1942, before the gathered commanders and construction experts of the western front, that the Atlantic Wall be built.

The Atlantic Wall was to consist of 15,000 bunkers, in which 300,000 soldiers would find protection from enemy fire, in order to be able to leave the protecting bunkers and fight against the enemy at the start of an actual landing.

At least they should be able to hold them on the beach until German reinforcements had arrived.

The keystone of the Atlantic Wall was to be the offensive long-range batteries at Cap Gris Nez. The OKW started with the assumption that the Allies would land on the coast between Ostende and Le Havre because of the short distance across the Channel, in order to make a direct drive toward the Ruhr region, the heart of the German economy.

Responsible for the building of the Atlantic Wall were the Todt Organization and its subcontractors, most of them local building firms, as well as naval fortress construction staffs. The Army and Navy had no command over the Todt Organization; they could only make recommendations to it. The completion of the Atlantic Wall was planned until May 1, 1943. More than 500,000 men were put to work, and all the available guns, even those from the West Wall and the Maginot Line, were put to use.

Despite the Todt's best ewfforts, though, the building of the Atlantic Wall moved forward slowly. Again and again, large contingents of workers had to be withdrawn to repair the damage done to Germany by Allied air attacks. All the same, in April 1943 the greatest monthly amount of 769,000 cubic meters of concrete had been erected.

In order to make the work of planning as simple as possible, the bunkers, rather like those of the West Wall, were assembled in a system of regular construction. This system was based on standardized bunkers, so that as soon as the decision of which type was needed was made, the work could start at once, since all the drawings and plans were already at hand.

Similar guns such as the French 15.5 cm K 418 (f) or K 420 (f) field cannons, Flak guns like the 88 or 105 mm types, antitank guns like the 75 or 88 mm, etc., could always be installed in these standard bunkers.

Exceptions to these standard bunkers were just a few special designs for ships' or railroad guns. In all, 10,206 bunkers were built on the Atlantic Wall according to this system.

The conceptions of the Army and Navy about the building of their coastal batteries were basically different. The Army considered the guns' readiness to fight at the moment of landing to be decisive and thus chose positions several kilometers away from the coast. They thus followed the principle of setting up the artillery behind the main battle line and having fire directed by advanced observers.

A meter-high concrete wall and tank obstacles protect a sector of the coast near Hendaye against tank landings. (BA)

The Navy chose their battery locations in the immediate vicinity of the coast, in order to fire on sea targets by direct aiming. The necessary protection of the guns was provided by concrete bunkers and steel armor plate. In addition, the actions of Army and Navy in coastal defense had been defined clearly in the Führer's Directive No. 40 in March 1942. According to this, the Navy alone was responsible for fighting sea targets, though it could, if need be, call in the Army's coastal batteries and the Luftwaffe's Flak batteries as well.

Waging the warfare on land was the responsibility of the Army, which could also call in the batteries of the other Wehrmacht branches.

Of the more than 300 Army and Navy batteries on the Atlantic Wall, the gun types came from five decades and included 28 caliber sizes from 7.5 to 40.6 centimeters. Along with German-made guns, many Russian, Czech, French and British guns were used.

Naturally, using these captured weapons resulted in problems in obtaining spare parts and ammunition. Shot tables and service manuals generally had to be worked out by the artillerymen themselves.

Along with the construction work, a huge propaganda campaign was carried on. As of the summer of 1943, newsreels and newspaper reports portrayed the Atlantic Wall as a rock-solid wall, consisting of hundreds of gun batteries and secured by extensive minefields and shore obstacles.

In the same year, the film "Der Atlantikwall entsteht" was made, commissioned by the Todt Organization.

All of this was meant to create—and not only in Germany—the impression that any landing on the West European coast was doomed to defeat from the start.

In actual fact, a tremendous construction job had been completed by the Todt Organization and the fortress engineering staffs before the Normandy invasion began. Over 8500 bunkers had been finished by that time. In spite of that, large portions of the 4000-kilometer coastline of the Atlantic Wall, which included the coasts of Holland, Belgium and France to the Spanish border, remained scarcely or not at all fortified at that time.

The coast of Normandy between the Vire and Orne rivers was not considered a possible landing site by the OKW because of the often rocky shoreline.

In this area there were, as of the spring of 1944, only a few Army batteries with French field cannons of 15.5 cm caliber. Beyond that, the shore was only protected by a few nests of resistance.

Above: Railroad Battery 674 near Hendaye was the southernmost battery of the Atlantic Wall. It was armed with three 24 cm Theodor cannons. (BA)

A 5 cm KWK L/60 in a concrete mount near St. Aubin sur Mer in Normandy.

These nests of resistance stretched along the coastline, mostly with a length of 300 and a depth of 200 meters.

The primary armament consisted of antitank guns in bunkers, either the 47 mm Fortress Pak (t) or the 50 mm Kampfwagenkanone (KWK). There were also captured tank turrets on ring mounts, field cannons, mortars and light Flak weapons.

All the parts of the facilities were linked by trenches. The outside security consisted of barbed wire, mine belts, tank traps and beach obstacles.

The crews, at platoon strength, were usually commanded by a lieutenant or sergeant.

In the nests of resistance, artillery observers of the Army coastal batteries often had their places as well, from which they were to direct the fire of their guns by radio or telephone.

In January 1944,. when Field Marshal Rommel took command of Army Group B, which included the coastal areas from the Netherlands to the mouth of the Loire, he immediately had additional batteries transferred to Normandy.

Two batteries of C/28 15 cm ships' cannons near Longues and Vasouy, plus a 21 cm K 39 Skoda battery near St. Marcouf were gained, and the Todt Organization also began to build 38 cm ships'-gun batteries near Le Havre and Cherbourg.

In addition, Rommel had shore obstacles set up on every bit of beach in his area where he considered a landing possible. These obstacles, most of which he designed personally, consisted of gatelike iron structures, tree trunks with metal spikes, and bent rails. The effect of the obstacles was often heightened by attached mines and grenades with contact fuses.

On the morning of June 6, 1944, the long-awaited Allied invasion took place, with 3.5 million soldiers, more than 20,000 aircraft, 3500 freight gliders and more than 5000 ships having been readied for the operation.

Four beachheads were situated between the Orne and Vire, the fifth on the peninsula of Contentin. The Germans had not considered the invasion sites to be endangered, [p. 7] and thus the Allies could soon break through the Atlantic Wall, which was weak there.

The Kora Battery (MAA 282) on the Ile de Re was equipped with two twin turrets from the cruiser Lützow. The 203 mm guns had a range of 37,000 meters and were among the most modern guns on the Atlantic Wall. After La Rochelle and the Ile de Re had been surrounded by Allied troops in August 1944, the battery often intervened successfully in land combat. (BA)

The German coastal defenses achieved only a few successes.

On the US landing sector of "Omaha Beach", the prepared Allied bomb carpet was fired at the hinterlands by a faulty calculation.

Thus the nests of resistance remained intact and could stop the US troops at the shoreline with concentrated defensive fire. Only after steady fire from battleships and destroyers did the resistance there break down toward evening.

Among the most successful shore batteries during the invasion was the naval battery at St. Marcouf. This battery, under the command of Oberleutnant Ohmsen, was able to sink two US destroyers. The battery also used its guns to fire on the nearby "Utah Beach" sector for days and make the landing of troops and materials much more difficult.

Despite the heaviest attacks by US troops, the battery held out for six days and was only abandoned after the destruction of all its guns. For this extraordinary achievement, the battery chief was awarded the Knight's Cross.

The Standard Bunker 221 for an 80 mm Grenade Launcher 34, without a covering of soil.

Below: An 80 mm Grenade Launcher 34 in Standard Bunker 221.

Below: The entrance to a minefield on the Channel coast.

In the 6 Scharten turrets, two machine guns on Scharten Mount 34 could be set up, as well as a parabolic telescope with fivefold enlargement and 14-degree field of vision around 360 degrees. These armored turrets, manufactured by Krupp, proved themselves very well in the fighting on the Atlantic Wall and withstood even the heaviest fire. The picture shows a 6 Scharten Turret at the entrance to Calais harbor.

The British coast called "Sword" had to be partially evacuated on June 26 because of continuing fire from German Army batteries, which had removed their guns from the bunkers to attain a greater field of fire. Of course, the Allies had meanwhile enlarged their other beachheads to the extent that this no longer mattered.

Seen in hindsight, the Atlantic Wall alone, without cover from the Luftwaffe and support by armored and infantry divisions, cound not hold back an invasion.

In spite of that, even at its weakest points it presented a considerable obstacle. The Allies were compelled to make extensive preparations for a landing on the coast.

Even today, one can see many of the bunkers of the former Atlantic Wall on the coasts of Holland and France. A few still contain guns. In the invasion areas in particular, bunkers and guns have been restored for museum purposes, and the visitor can thus get a good idea of what the Atlantic Wall once was.

A 155 mm Field Cannon 420 (f) of the Mont Canisy army battery. Despite heavy bomb attacks and fire from ships' artillery, this battery took the Allied "sword" beach under fire and thus compelled a partial evacuation of that sector. (BA)

5 cm Kampfwagenkanone 39

The 5 cm Kampfwagenkanone (KwK) in a makeshift pivot mount was used to fire on land and sea targets, and was used by all Wehrmacht branches on the Atlantic Wall.

The gun was developed and built from 1939 to 1941 by the Rheinmetall firm for the Panzerkampfwagen III (SdKfz 141) tank. In Panzer III Types E-H the 5 cm KwK L/42 was used, in Types J and L the 5 cn KwK L/60 was installed.

Of the 9568 5 cm guns that were made, over 1800 were assigned to coastal defense in the makeshift pivot mount (BhSkl) from 1942 on. The makeshift pivot mount consisted of a mount and a baseplate. The barrel with bottom section, barrel brake and pneumatic recuperator was screwed onto the cradle carriers. These, with the shield journals, were in the journal bearings.

To the right, at the edge of the mount, were the traverse and elevation aiming controls, as well as the seat and footrest for the aiming gunner. In the center of the mount was the socket of the turning column, with which the mount rested on the column.

The column in turn was welded to the baseplate by which the gun was attached to the concrete foundation. A chain was stretched around the outer extent of the baseplate and connected with the chain wheel of the traverse control. In the open ring mounting, a traverse field of 360 degrees was possible.

At the front of the mount, a shield of doubled armor plate could be attached to provide protection for the gun crew. The 5 cm KwK was aimed through the use of a 3 x 8 degree, later 1 x 11 degree, aiming telescope. The gun was fired electrically, firing the following shells at their listed initial velocities:

5 cm KwK	L/42	L/60
Explosive Shell 38	450 m/sec	550 m/sec
Antitank Shell 39	685 m/sec	835 m/sec
Antitank Shell 40	-	1190 m/sec

The maximum firing cadence was 15 rounds per minute.

Right: A 5 cm KwK L/60 of Resistance Nest 5 near La Madelaine in Normandy. The barrel is set at its maximum elevation angle of + 43 degrees.

Below: In this Standard Bunker 60, a 5 cm KwK L/42 has been installed. (BA)

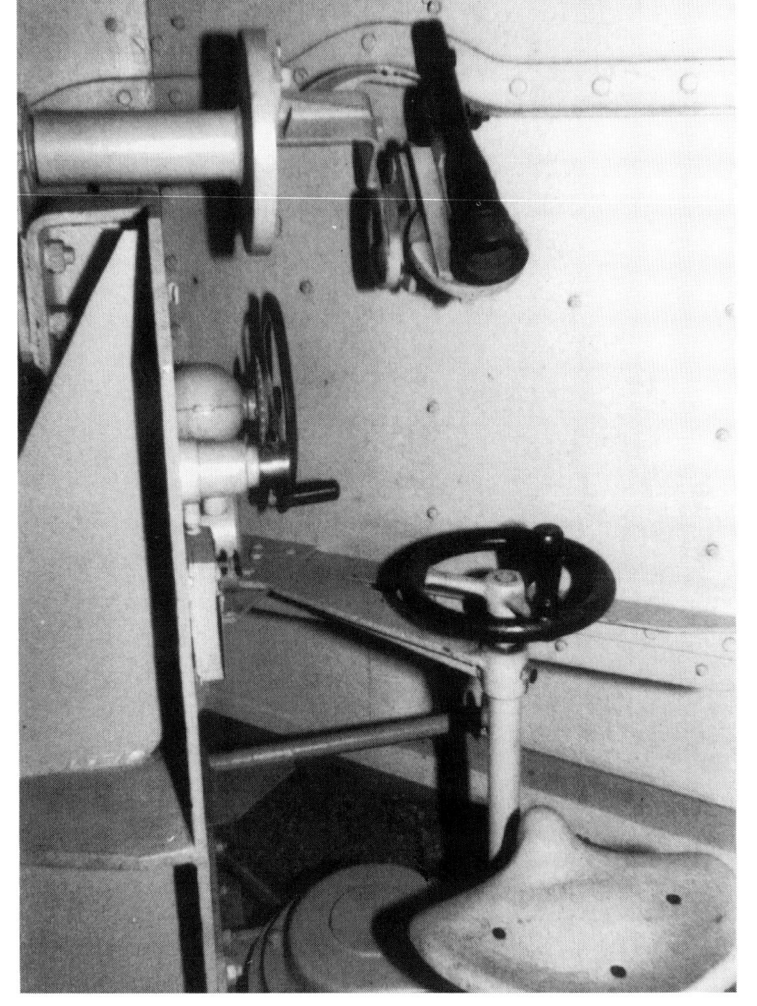

Above: This picture shows the lower part of the mount and the baseplate of the 5 cm KwK. The holes in the plate were used to fasten the gun to the bunker, and the channel for the traverse aiming chain is also welded on.

Left: The aiming gunner's seat on the right side of the mount. Above the seat is the handwheel for traverse aiming, and to the left is the elevation handwheel. To the right shield mount the aiming scope is attached; it turns along with the barrel. To its left is a distance drum with its handle. By turning the drum, the situation of the scope's line of vision with respect to the barrel is changed, and thus the aiming angle is established. For the antitank shell (numbers in black), distances of 200 to 1600 meters are settable, for the explsive shell (numbers in red), from 200 to 2100 meters. Under the scope is a scale with a handle for setting the side holders from 0 to 100 lines to left or right, divided into units of five lines.

Above: These two photos show a 5 cm KwK at Fort du Roule near Vherbourg. The double shield, made of three sheets of armor, is easy to see. On the right side of the shield is an opening for the targeting scope.

Below: This picture shows the welded design of the mount. Inside the mount is the turning pillar, which is welded fast to the baseplate. At right near the base is the hydraulic barrel brake cylinder.

Above: This 5 cm KwK L/42 stood on the promenade of the sea resort of La Baule. The artillerymen are just receiving things to read from a front library van. (BA)

Below: Alarm at a gun position. The camouflage nets have been taken down and the gunners are getting containers of six explosive shells ready. (BA)

Right: This KwK secured the harbor of Courseulles sur Mer. In the background, Allied invasion vessels lie at anchor.

Left: This gun took a direct hit on the muzzle brake and was thus put out of action.

Right: This 5 cm KwK L/42 also stood at Courseulles sur Mer; its shield was penetrated by two antitank shells, and part of the shield was torn away.

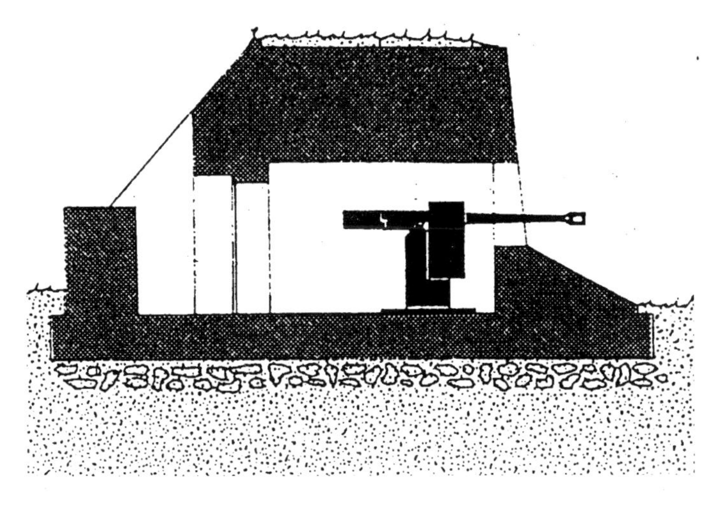

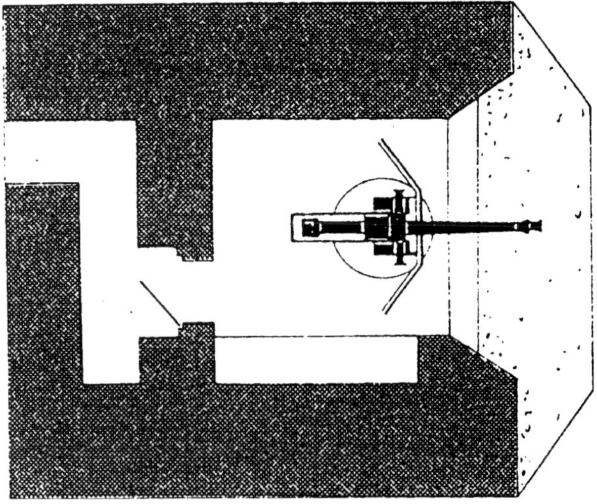

Above: Drawing of the smallest standard bunker 667 for the 5 cm KwK.

Left and below: The 5 cm KwK were often flanking, pointing alone the shoreline. On the ocean side, the bunker and gun barrel were protected from enemy naval fire. The upper photo shows standard bunker 667, the lower one standard bunker 613.

75 mm Field Cannon 231 (f)

The 75 mm field cannon came into being out of a request made by the French Army near the end of the last century. This called for a cannon that could fire at least fifteen rounds per minute. The designers of the Fonderie de Bourges firm in Bourget and Schneider et Cie in Le Creusot achieved this goal through the use, for the first time, of a hydraulic brake cylinder. It was filled with a mixture of oil and water and moved the barrel, recoiling after a shot, back to its firing position immediately.

This returning method, very protective to the mount, callowed a shot cadence of eighteen rounds per minute.

The cannon was introduced into the French Army in great numbers under the designation "Canon de 75 mle 1897".

During World War I, the 75 saw action on all fronts, with great success; it was regarded by the French as a siege instrument, and one could even buy postcards and pendants depicting the gun. Not only did it prove itself very well as a field cannon, but it was also able to gain success as a makeshift anti-aircraft gun with a raised mount.

After the end of World War I, the 75 mm cannon remained in service with the artillery, and was also taken over by the US Army as its standard cannon.

In the thirties, the guns were fitted with modern rubber tires in place of its iron-rimmed wooden wheels. In World War II, the 75 mm cannon was again used against German troops until the French campaign ends, when it was taken into the possession of the Wehrmacht.

Its German designation was "7.5 cm Feldkanone (FK) 231 (f)", or also 7.5 cm FK 97 (f).

At the beginning of the Russian campaign in 1941, the 75 mm field cannon was used briefly as an antitank gun, and then found its niche in the coastal defenses of the Atlantic Wall. Here it was used by the Army and Navy alike. Set up either in an open ring position or in standard bunkers 611, 612 and 662, the cannon could be used effectively to provide barrage fire. It fired explosive shells with a weight of 5.5 kilograms and a range of 7800 meters, and antitank shells, weight 7.4 kg, range 8500 km. This ammunition all came from old French production and thus often proved to be faulty.

In 1944, the French Army recaptured 75 mm cannons and used them again, and after World War II the cannons, after almost fifty years, found their final homse in museums.

The weight of the 75 mm Field Cannon 231 (f) in firing position was 1140 kg. The range of traverse was three degrees to either side, and the rate of elevation from 11 to 18 degrees.

This 75 mm Field Cannon 231 (f) belonged to the Army's Coast Artillery Unit 1280 and was stationed on the Ile de Re in the west. A practice firing is taking place to be photographed. The aiming gunner behind the aiming optics is sighting a target, the loading gunner already has the trigger cord in his hand and is waiting for the order to fire. The gun leader is watching the practice from the right. (BA)

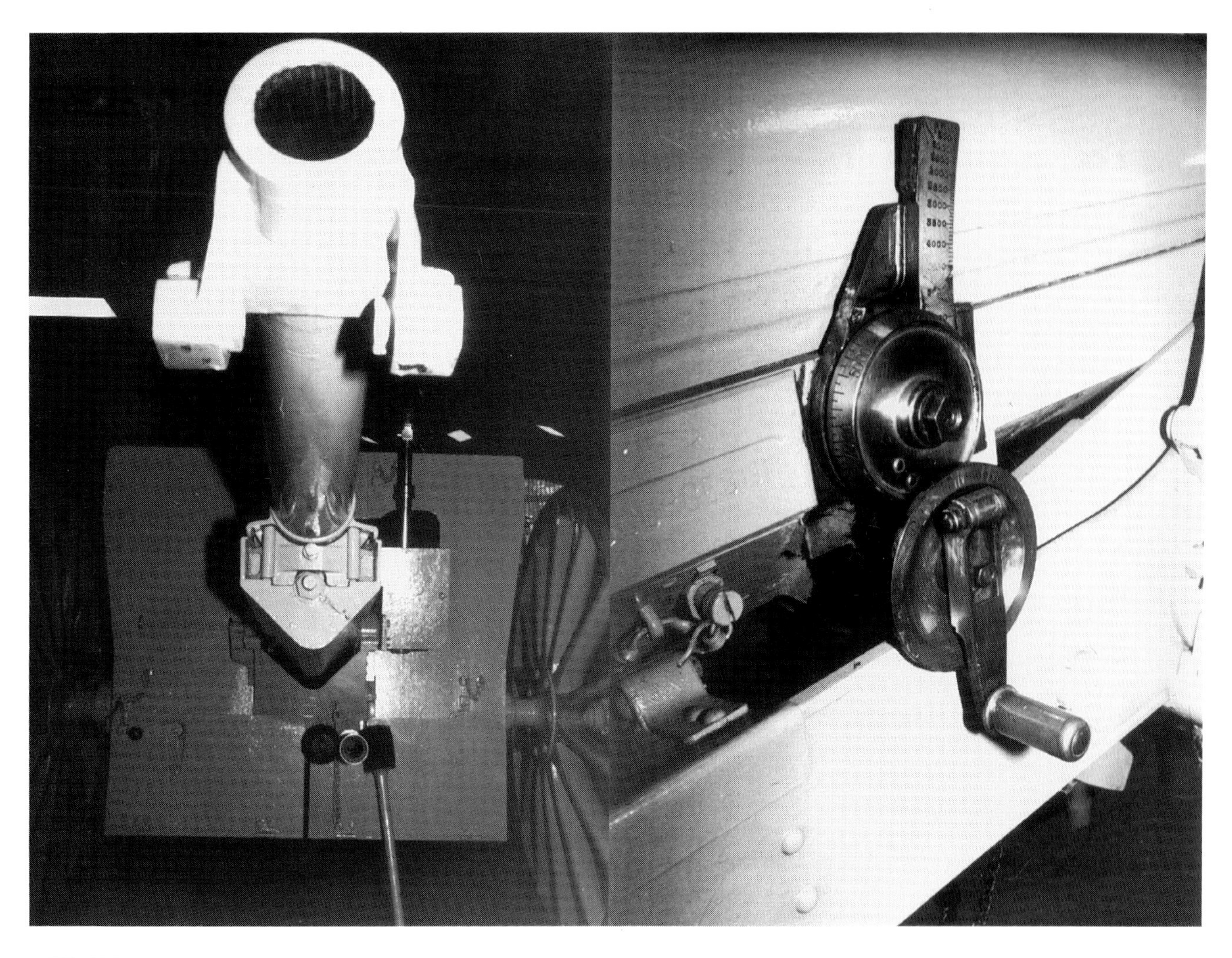

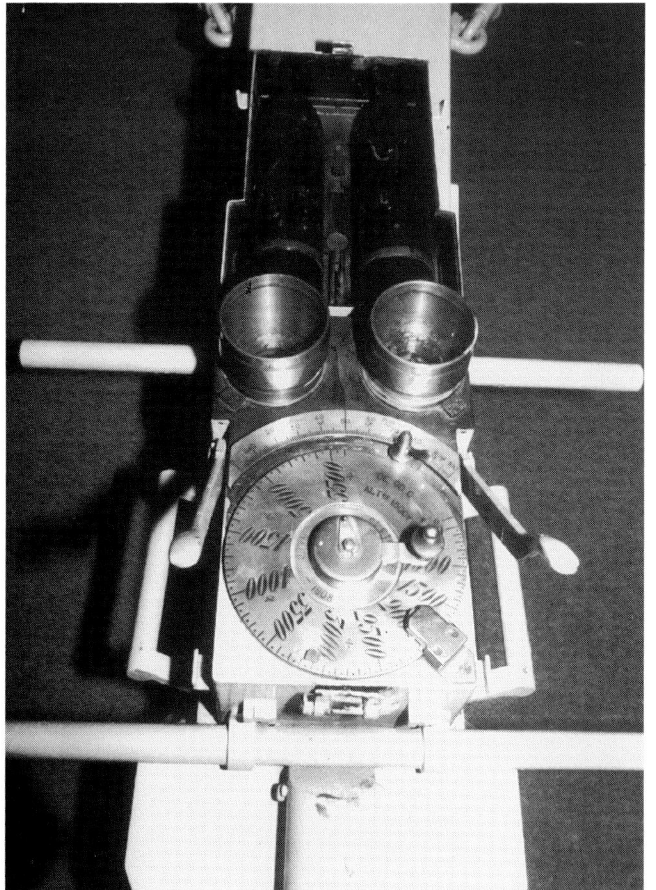

Upper left: A look at the barrel, 2590 mm long, of the 75 mm cannon. The gun had only a simple shield of armor plate.

Above: The elevating crank of the 75 mm cannon, with which the aiming gunner could read the set height on the upper wheel. At the same time, the height was indicated on the scale above.

Left: The fuse-setting machine with two cups, called "debouchoir" in the French directions. The scale reached from 500 to 5500 meters. The shells were held in the cups with the fuses, and by pressing the setting lever, the desired distance was set on the fuses.

Right: A look at the breech of the 75 mm cannon, with the traversing crank seen beside the barrel. With it, the barrel could be swung three degrees to either side. The seats for the aiming gunners on either side of the barrel were left off for coastal servies, since they were only a nuisance.

Left: A 75 mm cannon in the Army's Merville Battery in Normandy. In the night of the invasion, the battery was captured by British paratroopers and the cannons were destroyed. Beside the gun lie several empty ammunition cases.

Right: Two guns of the Adour South Battery at the mouth of the Adour, in makeshift positions. (BA)

The 75 mm cannon was also used as make-shift anti-aircraft weapons in Navy batteries. Here a Flak battery fires a test shot near Biarritz. (BA)

Right: The pillar for the 75 mm cannon.

Left: A 75 mm cannon at the Bt. de Gavres Battery near Lorient. The gun is protected by sandbags in a make-shift position. (BA)

88 mm Ship's Cannon C/35

The 88 mm Ship's Cannon C/35 in the U-boat Mount C/35 belonged to the artillery weapons of the German U-boat Types VII A to D. As of May 1942, more and more of the U-boat crews dispensed with these weapons, since above-water attacks had become as good as impossible becase of Allied defenses. Combined in batteries of from three to six cannons, they strengthened the weaponry of the naval artillery units, usually in the vicinity of the U-boat harbors on the Atlantic. The caliber was 88 mm, the barrel length 3450 mm. The range of traverse was 360 degrees, reduced to 120 degrees when installed in standard bunker 670 or 671. The range of elevation extended from -4 to +30 degrees.

The guns fired the Type 39 antitank shell and the L/45 explosive shell with head fuse. The muzzle velocity was 700 meters per second, the maximum shot range 12,350 meters. The cannons were made by the firm of Rheinmetall-Borsing AG in Düsseldorf.

An 88 mm ship's cannon at the entrance to the harbor of Brest.

Below: An 88 mm cannon in standard bunker 671. The handwheel of the elevating mechanism can be seen below at the left, beside it that handwheel for the traversing mechanism. Above the barrel the targeting apparatus is attached; this moves along with the elevation of the barrel. (BA)

The U-boat Mount C/35; this gun is presently in the Defense Technology Study Collection in Koblenz. The handwheel for the traversing mechanism has been removed.

A look at the two barrel-brake cylinders and the gearing for the elevating mechanism between them. The bolt in front of the geared arc was pushed down at the zero position of the barrel and thus stopped the gun's motion.

Below: A rare picture taken during the firing of an 88 mm SK. The aiming gunner sights the target through the aiming optics, while the traversing gunner stands at the traversing handwheel beside him. Behind the gun, explosive shells are handed in by the ammunition gunners.

94 mm Anti-Aircraft Gun Vickers M (e)

The 94 mm anti-aircraft gun was developed and built in 1939 by the Vickers firm and the Elswick Ordnance Company.

The gun was the standard British AA gun and was used with great success against the German Luftwaffe and, as of 1944, also against the V 1 rocket.

In 1940 the Wehrmacht captured a series of these guns in Norway and in France, near Dunkerque.

These guns were given the designation "9.4 cm Flugabwehrkanone Vickers M 39 (e)" and used by naval artillery units as anti-aircraft guns, and later as light sea-target artillery.

The attacking of air targets waas done by direct aiming, with German-made line sights, later with optics.

The two aiming scopes were attached separately on the left and right sides of the gun, and were adjusted parallel for elevation and traverse.

As of 1942 the 94 mm Vickers saw service only against sea targets because of a growing shortage of ammunition.

Right: Two 94 mm Vickers M 39 (e) guns of the Creche II Battery north of Boulogne. (BA)

Below: The Le Portel Battery, south of Boulogne, was likewise equipped with the Vickers guns. In the summer of 1940, the guns were still in open firing positions; later three Type 671 bunkers were built. (BA)

Upper left: Front view of the 94 mm Vickers mount. Bext to the geared arc for barrel elevation, the receivers for the elevation and traverse angles are visible.

Above: The centrally located breech of the Vickers AA gun. At the left side is the folded-up loading tray, at right the barrel balance weight.

Left: The barrel of the 94 mm Vickers gun with its balance weight and the barrel brake cylinder on top of the barrel.

Two pictures of the Le Portel Battery.

Right: Before the 94 mm mount stands the gun leader, Feldwebel Zeidler. At the left edge of the picture, the traversing mechanism can be seen.

Below: Gun practice. The two traversing gunners operate the traversing mechanism. The ammunition gunner stands on the loading platform and has already inserted a shell. Before him stands the loader with the trigger cord. The gun leader oversees the drill.

Alarm at the Creche II Battery. At the right side of the gun, the aiming gunner is viewing the target with the line scope. This was made by the naval artillerymen themselves after the British had destroyed all the targeting apparatus for the 94 mm Vickers guns before they withdrew. (BA)

Below: Ammunition storage in the gun position.

10.5 cm Field Cannon 331 (f)

The 105 mm field cannon was produced by the Schneider firm in 1913 and introduced into the French Army in the same year. It too saw action on all fronts in World War I, and remained in service after the end of the war. The life story of this field cannon is much like that of its 75 mm counterpart. In 1940 the field cannon was also taken over by the Germans and designated 10.5 cm Feldkanone 331 (f). Because of its age and the usually very worn condition of its barrels, the gun was soon assigned to coastal defenses.

Here they were put to use by the Army and Navy, at first in field positions, later the standard bunkers 649, 650, 651, 652, 669 and 670. In the bunkers the wheeled mounts, hard to aim, were often removed and the barrel mounted on a German naval gun pillar made around the turn of the century. The range of traverse of the 10.5 cm Field Cannon 331 (f) in the bunker was reduced to 120 degrees, and the shield, which turned along with the gun, guaranteed protection to the gun crew.

Despite their age, the 10.5 cm field cannons were used in goodly numbers along the Atlantic Wall and proved themselves well during the invasion and the subesquent combat.

This cannon too can still be found in the museums and bunkers of the Atlantic Wall.

The weight of the cannon with its wheeled mount, in firing position, was 3300 kilograms, of which the barrel, with a length of 3820 mm, weighed 1105 kg. The geared aiming machinery allowed a barrel elevation from -10 to +18 degrees. With a shot weight of 15.74 kg, the shot, with a muzzle velocity of 550 m/sec, attained a maximum range of 12,000 meters. In practice, though, the used barrels often allowed only a range of 9000 meters.

The 10.5 cm Field Cannon 331 (f) seen from the front. Under the barrel is the access hatch for the barrel brake cylinder. The hand crank on the shield was used to stop the axle from turning.

The photo below shows the base of the gun with the breech of the Field Cannon 331 (f). At the left of the breech is the breech lever, in the middle is the trigger bolt, which was activated by a trigger cord.

Below: A 105 mm field cannon in a bunker at Corbiere Point on the Isle of Jersey. (Zimmermann)

Left: A 10.5 cm Field Cannon 331 (f) on a turn-of-the-century naval mount. (Zimmermann)

Right: Camouflaged bunkers of the 1./ 1280 Army Coast Artillery Regiment near Les Sables d'Olonne, with a 105 mm field cannon.

Left: The massive bunker offered gun and crew sufficient protection. Wooden planks have been installed around the opening to catch shell fragments. (Zimmermann)

15.2 cm Cannon Howitzer 433/l.r

In the early thirties, the Soviet Union began the planned development and production of modern cannons.

Numerous models came into being then, including the 152 mm Gaubitsa-Pushka obr. 137 g (ML-20) Cannon-Howitzer.

This gun was built in very large numbers from 1935 to the end of World War II.

As of 1941, the cannons were captured by the Wehrmacht on the eastern front, and put to use with the designation of 15.2 cm Kanonenhaubitze 433/1 (r).

In 1943 the cannon howitzer was likewise turned over to the Army Coast Artillery because of ammunition shortages.

The exact caliber was 152.4 mm, the barrel length 4405 mm (L/29), with the rifled part 3476 mm long.

The mount allowed a traverse field of 58 degrees and an elevation field from -2 to +65 degrees.

The total weight of the cannon howitzer in firing position was 7128 kilograms.

The antitank and explosive shells were loaded separately from the cartridge and weighed 43.5 kg. With a muzzle velocity of 655 m/sec, the maximum range of 17,265 meters could be attained.

Right: The barrel of the 15,2 cm Cannon Howitzer 433/1 (r) with its muzzle brake.

Below: Work on the breech of this cannon howitzer, in an open ring bed, is being carried out. (BA)

Left: The base and breech of the cannon howitzer. On the right side of the shield is the container to hold the aiming optics during transport. In firing position, this was attached to the left side of the shield behind the visor flap.

Below: The muzzle brake with its twelve side louvers on either side braked the recoil of the gun when firing by means of a flange, arched toward the back, in front of every opening. In the moment when the shell passed the muzzle brake and thus closed it off, the powder gases were forced out the side louvers and directed backward. Thus the recoil of the barrel was braked and partly eliminated.

Right: The wheels of the cannon howitzer were rubber-tired. Above the wheels, the shield attachments and shock absorbers can be seen.

Below: The front part of the cannon howitzer, the shield having been removed here. Under the shock absorber is the traverse gear, which gave a traverse range of 29 degrees on either side.

Below: In the bottom picture, the gun is being moved into the open ring position by the strength of its crew. (BA)

These two photos show the Chiberta Battery near Biarritz. The battery consisted of six Cannon Howitzer 433/1 (r) guns in open ring beds. In the upper picture, a gun is just being aimed crudely. (BA)

In the lower photo, a lesson in cartridges is just being given. (BA)

15.5 cm Field Cannon 418 (f)

The field cannon with the designation Canon de 155 GPF (Grand Puissance Filloux) was put into service by the French Army in 1917. One year later it was likewise taken over by the US Army as 155 mm and M 1918 M 1 by the field artillery, and it remained in service until the beginning of World War II.

In 1940 there were still 449 of the 15.5 cm GPF cannons in use by the French Army, and after the armistice the greater portion of them were taken over by the Wehrmacht as the 15.5 cm Feldkanone 418 (f).

From 1942 on, the 15.5 cm Feldkanone 418 (f) strengthened the coastal defenses in Army coast artillery units.

The barrel length was 5915 mm, of which the rifled section was 4483 mm long; the mount allowed a traverse range of 30 degrees each way and an elevation range from 0 to +35 degrees.

The gun fired explosive grenades with a muzzle velosity of 735 m/sec. The maximum shot range was 19,500 meters.

Right: This Field Cannon 418 (f) is in position, with its barrel at its highest elevation of +35 degrees.

Below: The Socoa West Battery (2/MAA 286) near St. Jean de Luz wasa equipped with four of the Field Cannon 418 (f). The guns were in standard bunkers of Type M170 or H622, and had been remounted on turning mounts from naval stocks. (BA)

This Field Cannon 418 (f) stands in an open field position. The pair of front wheels were pushed between the mount spars.

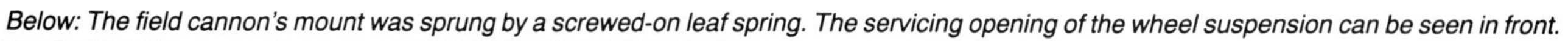

Below: The field cannon's mount was sprung by a screwed-on leaf spring. The servicing opening of the wheel suspension can be seen in front.

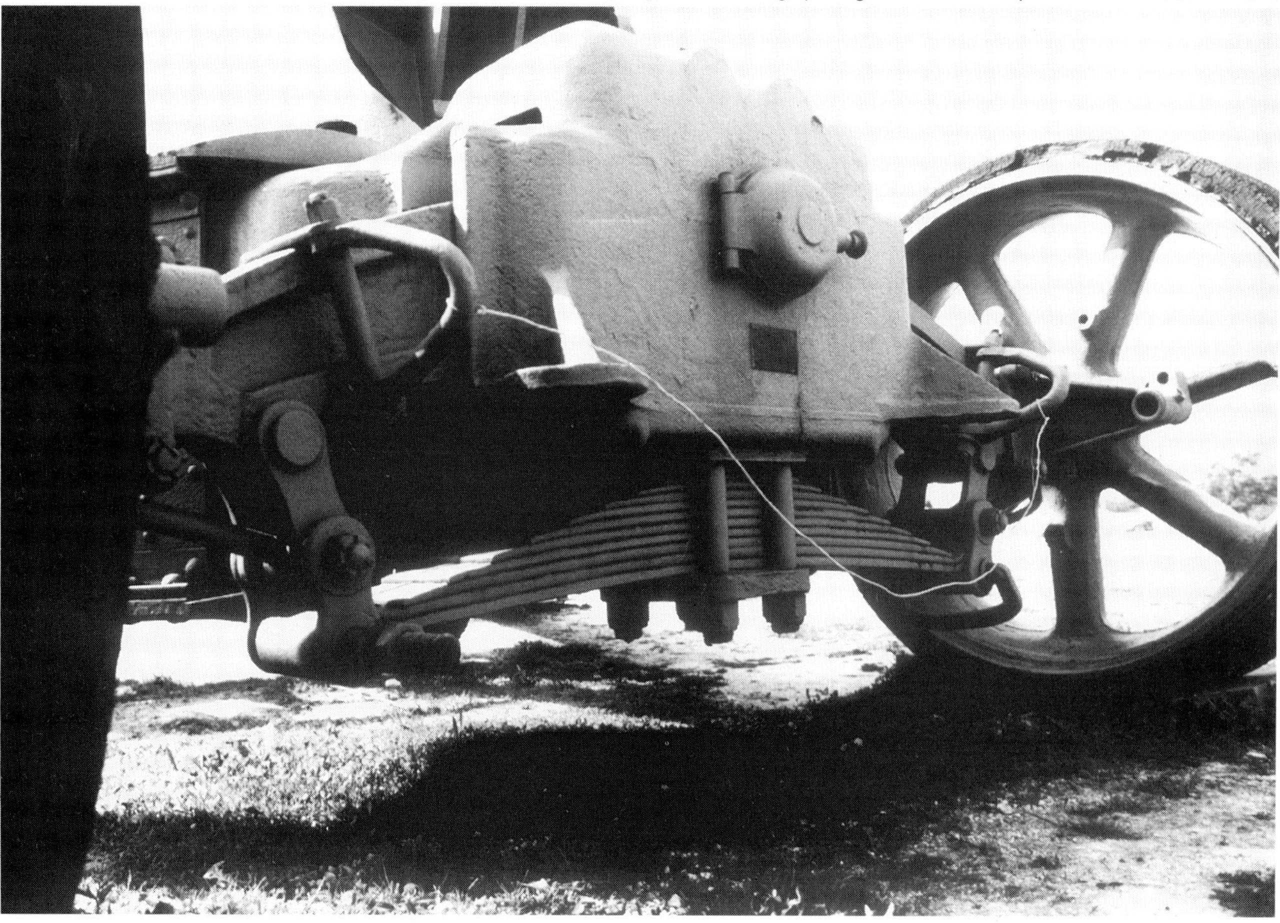

Right: At left in the photo, thr aiming gunner is sighting a target, while the elevation gunner next to him is cranking the barrel to the ordered height. (BA)

Below: This 15.5 cm Field Cannon 418 (f) of an Army coastal artillery unit is just being settled in a new firing position. The wheeled mount is attached to a turning plate, which makes a quick traverse setting of the gun possible. (BA)

Left: Setting the traverse of the field cannon was hard work, what with the gun weighing 10,750 kilograms. The traversing gunners had to push the gun to the ordered traverse setting with a rod. Here the degree numbers have been painted on the surface of the gun bed. (BA)

Lower left: The mount of the Field Cannon 414 (f) with the shield attachment, photographed from the side.

Below: The mount spars have been fixed in position with iron rods and posts in the ground.

A camouflaged gun position with a Field Cannon 418 (f) on the Belgian coast. (BA)

Left: A blown-up Field Cannon 418 (f) of the Moltke Battery on the Isle of Jersey. The command post can be seen in the background. (Zimmermann)

Right: A bunker for a Field Cannon 418 (f) at the Point du Hoc, the bunker showing heavy shot damage from Allied bombing attacks and naval artillery fire.

15.5 cm Field Cannon 418 (f) in Open Ring Bed

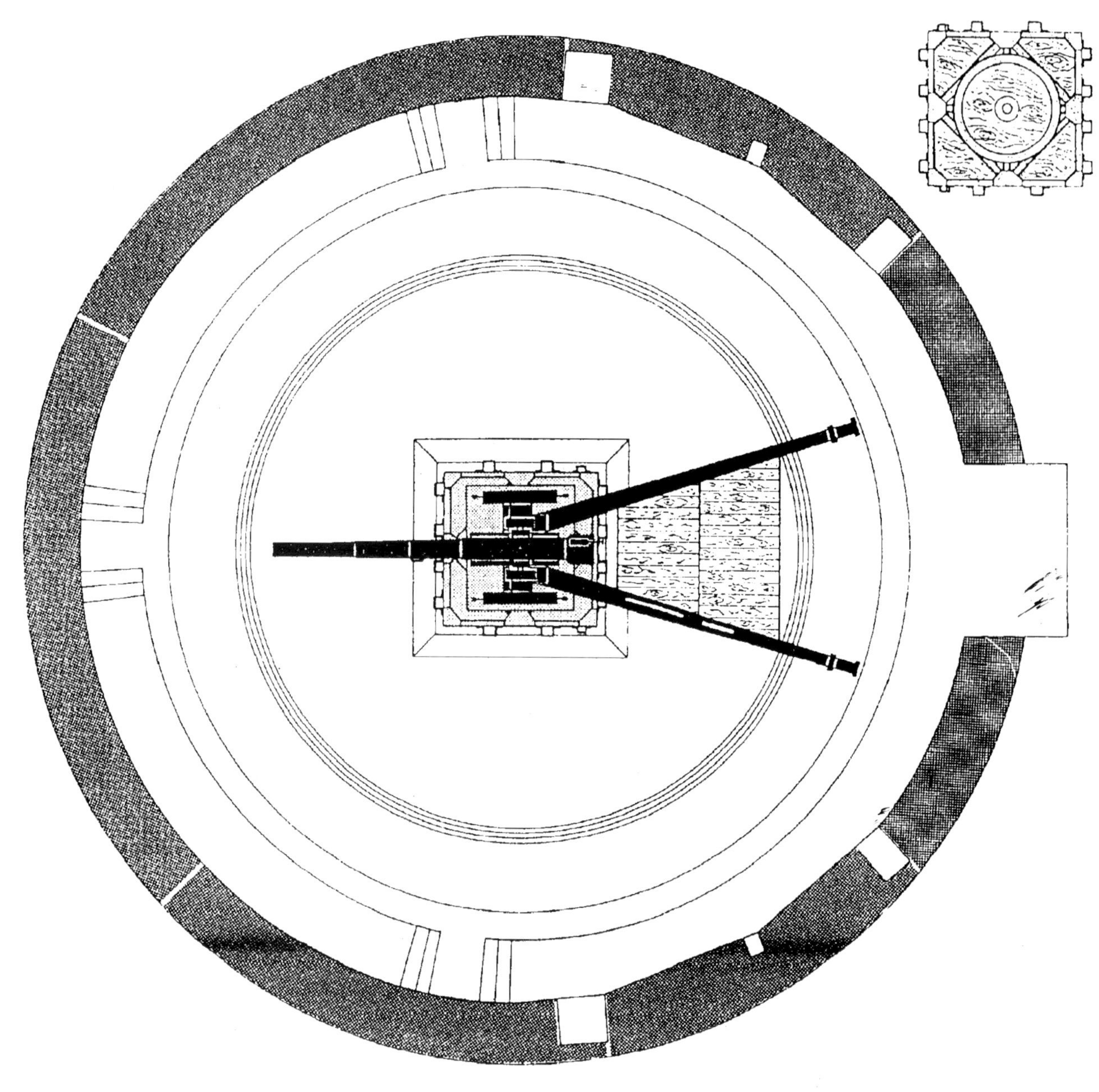

23-3 L/82

B 37 0B

28 cm Railroad Cannon 5

In the summer of 1940, three railroad batteries, each with two 28 cm K 5 guns, were transferred to the Channel coast by the Wehrmacht for Operation Sealion.

Battery E 712 took up a position near Pointe aux Oies, E 713 near Hydrequent and E 765 at first in the railroad yards of the harbor depot of Calais; later this battery was transferred to Coquelles.

After Operation Sealion had been given up, the three batteries remained on the Channel. To protect the cannons, the Todt Organization built two domed bunkers for Batteries 712 and 713, with room inside for the guns and their Diesel locomotives. For Battery 765, a multistory tunnel layout was built in a quarry near Coquelles. This also contained ammunition stores, housing and administrative offices. The K 5 guns were aimed via a built-in rail-switching system and turntables.

Along with the Navy's long-range guns, the 28 cm railroad guns dominated the Channel and the south coast of Britain.

In addition, they could fire on any shipping traffic at the harbors of Ramsgate, Dover and Folkstone.

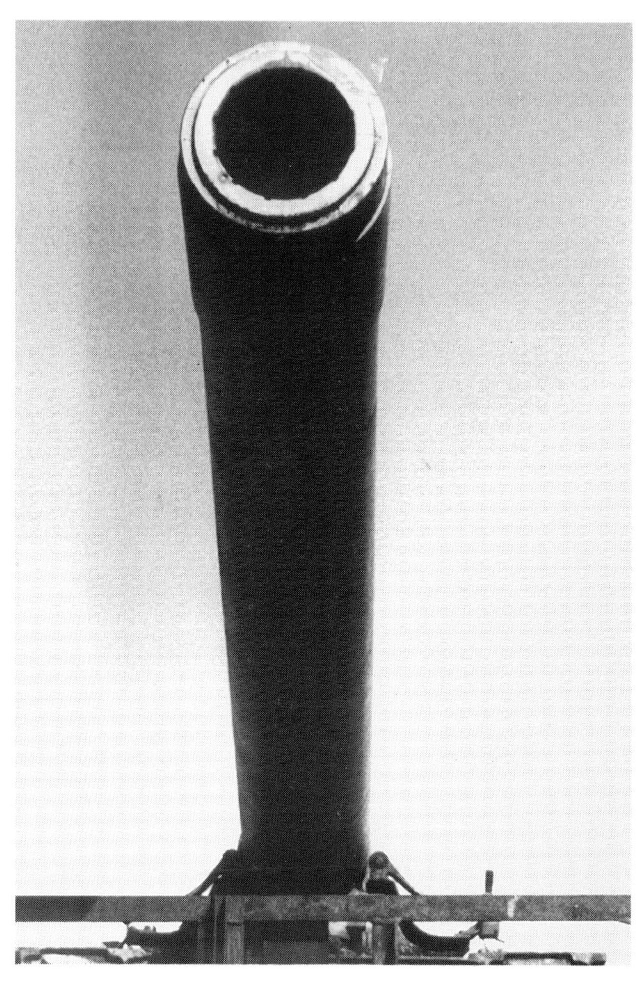

Above: The mouth of the barrel of the K 5, over 21 meters long, with twelve riflings. The inner barrel can be seen protruding slightly from its mantle.

In the photo below, a group of Japanese visitors is examining Railroad Battery 765 in the harbor of Calais. (BA)

The rear six-axle running gear of the K 5. This was driven by a Diesel engine mounted above the wheels, so that the gun could move short distances independently without the help of a locomotive.

Below: In July 1992 the Atlantic Wall Museum near Audinghen was given a K 5 by the French Army. Here the barrel is being lifted onto its cradle by two special cranes.

For firing on the British coast, special shooting-in shells that left a visible trail of black smoke for observation could also be used. In all, the Krupp firm of Essen produced 25 of these K 5 guns, which weighed 218 tons apiece.

Left and below: These two photos show the ammunition crane of the K 5 gun. It was electric, but could, if necessary, also be operated by hand.

After the Allied invasion of 1944, the K 5 guns took part in the combat against the Canadian troops who advanced along the Channel coast.

The railroad batteries were then withdrawn in the direction of Holland early in September, and were later lost in the combat in Germany.

The 28 cm K 5 (E) was a profiled [?] railroad gun mounted on two six-axle trucks. The gun's design included a mantle and a changeable inner barrel.

The barrel, 21,539 mm long (L/76], was held by a special cradle with lengthening carriers.

The barrel's recoil of, at most, 1150 mm when fired was sprung by two recoil cylinders, and a hydropneumatic recuperator brought the barrel back to its firing position.

The K 5 barrels were built in various types: the K 5 deep-rifled with 10 m, K 5 multirifled with a conical barrel, and K 5 smooth with 310 mm caliber.

For the smooth barrel, fin-stabilized undercaliber shells with ranges up to 160 km; for the other barrels, shells with auxiliary rocket drive, with ranges up to 86,500 meters, were developed.

In the K 5 batteries E 711 and 765 on the Channel, only the deep-rifled barrles for standard shells were used. The explosive shells weighed 255.5 kilograms.

The Explosive Shell 35 with base fuse / 35 K, later also double fuse Z 45 K, with 125 seconds of running time, was fired aith a muzzle velocity of 1120 m.sec.

Above: *Here the shield attachments for the gun cradle are being installec*

Left: *Cartridge cases for the 60.5-kg main cartridge of the K 5.*

Below: *Under the gun cradle, the hydropneumatic recuperator, which brou the barrel back into its firing position after a shot, can be seen.*

K5 Batteries on the Atlantic Wall

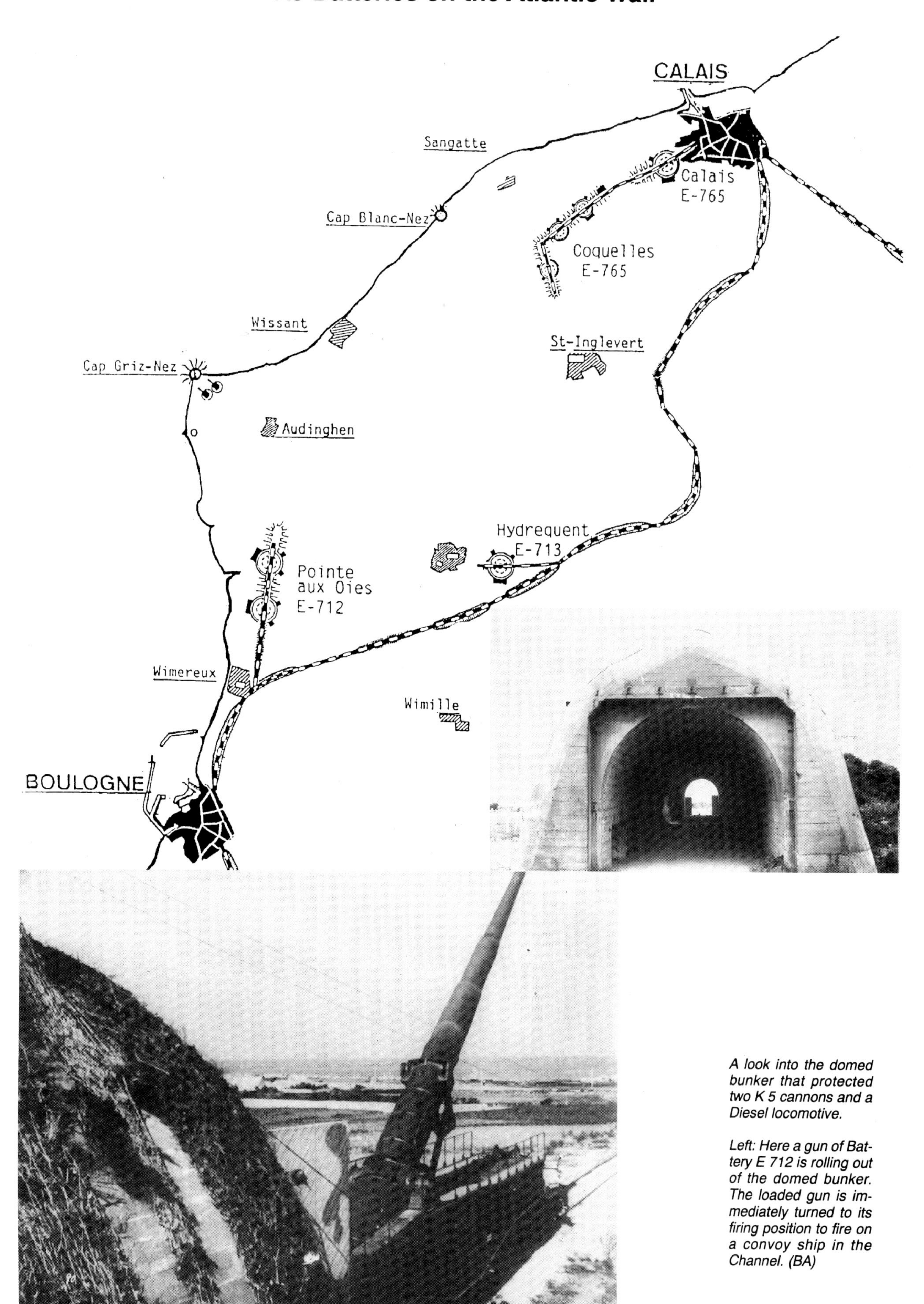

A look into the domed bunker that protected two K 5 cannons and a Diesel locomotive.

Left: Here a gun of Battery E 712 is rolling out of the domed bunker. The loaded gun is immediately turned to its firing position to fire on a convoy ship in the Channel. (BA)

A domed bunker of Battery E 712, with soil banked against it. The arched form of the bunker's roof was meant to deflect striking shells off to the side.

The drawing shows the great size of the bunker.

Below: Here a gun of Battery E 713 rolls out of its bunker near Hydrequent. The artillerymen have found seats on the gun. (BA)

Front view of the domed bunker for the K 5 gun.

Below: A K 5 of Battery E 713 has rolled onto the turntable, which could turn it 360 degrees. The gun is being prepared for firing. Two empty cartridges lie ready at the left side of the picture.

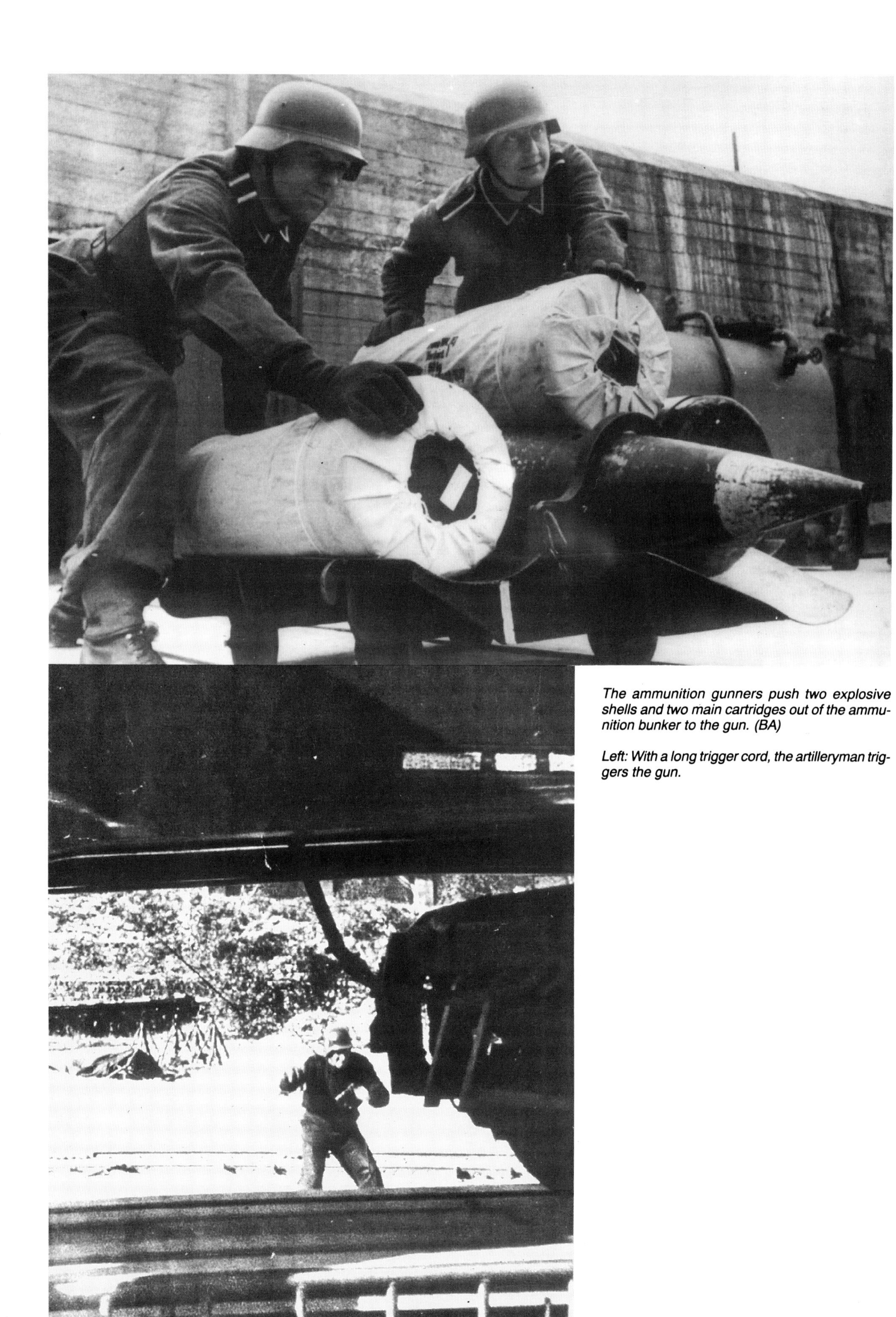

The ammunition gunners push two explosive shells and two main cartridges out of the ammunition bunker to the gun. (BA)

Left: With a long trigger cord, the artilleryman triggers the gun.

24 cm Railroad Cannon 675 (f)

The 34 cm railroad cannons were made in France in 1912, and were built for the French Navy by the firms of St. Chamond and Batignolles.

Four units, designated “Canon de 340 mm M Model 1912”, with a weight of 270 tons, were put into service by the French Navy in 1914, divided between two batteries.

The guns were first used against German troops in Flanders in the autumn of 1914.

As of May 1916, the 34 cm guns saw service on the Italian-Austrian front in the many battles on the Isonzo.

Their last use during World War I was on the Somme front, against the 18th German Army, which had pushed to within 85 kilometers of Paris in their spring offensive in March 1918.

After World War I, the 34 cm railroad gun batteries remained in service and were used as the 1st and 2nd Batteries of the 373rd Regiment to strengthen the Maginot Line.

In the spring of 1940 the batteries were at Rittershoffen and Lindel in Alsace, and were aimed at Karlsruhe and Pirmasens.

In the night of June 13-14, 1940 the French moved both batteries, without having fired a shot, in the direction of Belfort.

There all the guns were captured undamaged by German troops and transported to the Krupp works in Essen. Here the guns were given their final designation: 34 cm Eisenbahnkanone 675 (f).

The guns were issued to the newly created Naval Artillery Unit 264.

This unit, along with five Flak units, was assigned to the Atlantic seaport of Lorient, the main support point of the German U-boats, as well as the seat of the U-boat commander, in order to defend the fortifications.

The unit had five batteries, of which the one at Plouharnel, at the north end of the Quiberon Peninsula, had the largest-caliber guns, the 34 cm 675 (f) Railroad Guns.

After thorough ballistic testing by Krupp and the Navy, it was decided to keep the French mounts, which allowed an elevation from -8 to +42 degrees.

This was sufficient for use as coastal defense weapons, and thus Krupp only needed to build fixed turntables with a traverse field of 240 degrees.

In addition, Krupp built new loading cranes, which lifted the shells and cartridges onto the 3.4-meter-high loading platforms.

In the spring of 1942 this work was finished and three guns and turntable mounts were transported to Plouharnel, in order to be built into the beds that had been prepared for them.

The fate of the fourth gun is unknown.

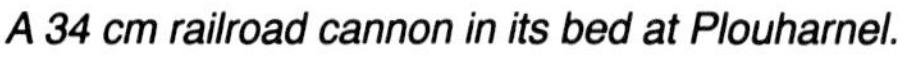
A 34 cm railroad cannon in its bed at Plouharnel.

The barrel of the 34 cm Kanone 675 (f) at its greatest elevation of +42 degrees. Camouflage nets have been drawn over the bunker and the rear part of the gun.

Every gun's bed had a diameter of 32 meters and was equipped with two bunkers for shells and cartridges.

The entire battery grounds spread over several square kilometers and were among the "finest structures" of the Atlantic Wall.

The battery was equipped with a high command post, several optical range finders, as well as a FuMo 2124 "Würzburg Giant" radar device.

The battery possessed its own source of energy, a sickbay and several reserve ammunition bunkers. The transport of materials and ammunition within the battery area was done on a narrow-gauge railroad line.

To secure the battery from outside, several kilometers of barbed-wire obstacles and tank traps were erected, while minefields and antitank guns secured the area for some distance.

Anti-aircraft protection was provided by its own 2 cm Flak 28 Oerlikon and 4 cm Flak Bofors M 1 guns.

The battery chiefs were, in order, Kapitänleutnant von Natzmer, Kptlt. (MA) Clages and Oberlt. (MA) Suling. The battery crews consisted of 310 artillerymen.

For foreign visitors from friendly nations to this part of the Atlantic Wall, a visit to the Plouharnel Battery was obligatory, and many Japanese naval officers, whose U-boats docked at Lorient, inspected the battery.

After the surrounding of the Lorient fortifications by Allied troops in August 1944, work began to turn the guns, which could only fire in the direction of the sea until then, to fire to the landward side. This laborious work was finished in January 1945.

To check the firing situation, Oberleutnant Suling had a salvo fired at the railroad station in Vannes 25 kilometers away. This important supply depot for the encircling troops was hit hard.

The good targeting was confirmed to the battery by a French official in Vannes.

From then on, the battery often intervened in the ground combat and destroyed several enemy battery positions. But it in turn was under heavy fire, and by the end of March, all the guns were so badly damaged that it was no longer possible to fire them; thus the battery was vacated.

After the war, French engineers blew up the guns, and what remained was cut up by scrap dealers.

The Japanese emissary Oshima visits the Plouharnel Battery. The Battery Chief, Kapitänleutnant (MA) Clages, is explaining the aiming of the guns.

Another photo of the aiming of a gun. At the right front, two men are operating the elevating crank; near them the meter-elevating arc can be seen. The traversing gunners, in the center of the picture, have already set the aim and wait for further commands. Up on the gun platform, the ammunition loaders await the order to fire.

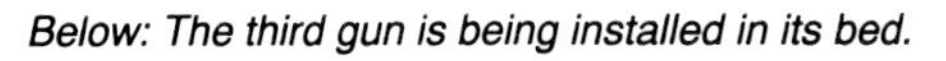

Below: The third gun is being installed in its bed.

Ammunition gunners are bringing two BM 16 charges of powder to the ammunition crane on a handcart. The barrel has already been placed in its loading position.

Below: Lettering on the wall of the ammunition bunker: And if the enemy flings iron and lightning, we'll stand fast by our gun!

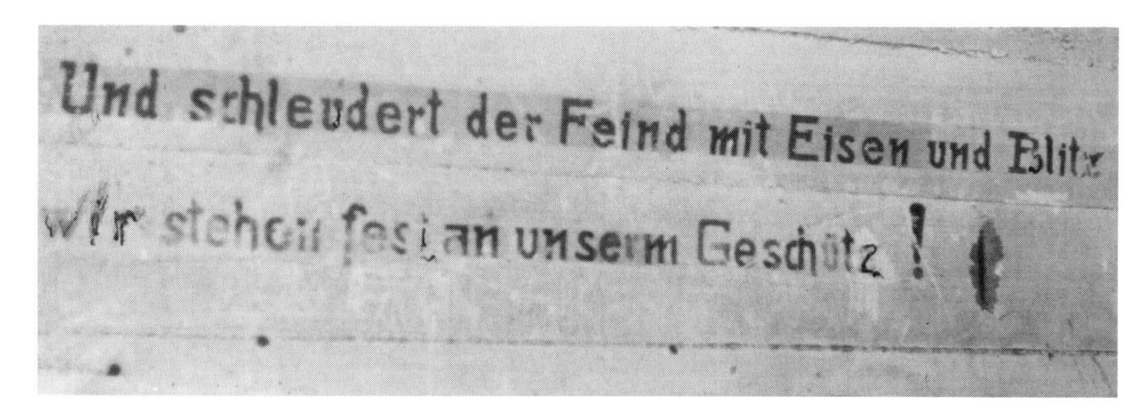

With the 34 cm cannon, the following shells could be fired. Depending on the maximum shot range desired, the quantity of powder could be reduced by using smaller charges.

Explosive shell (c.1915), weight 465 kg, length 1244.5 mm. With two charges of BM 16 powder, weighing 153 kg in all, the maximum shot range of 31,300 meters at a muzzle velocity of 850 meters per second was attained.

With two charges of BM 16 powder weighing 132.5 kg, a range of 25,000 could be attained, with a muzzle velocity of 750 m/sec.

Hooded shell (c/1912), weight 540 kg, length 1078 mm. The two propellant charges consisted of 152 kg of BM 17 powder, the range was 26,400 meters, with a muzzle velocity of 800 m/sec.

The hooded shell could also be fired with 106 kg of BM 15 powder; then the mazimum shot range was 20,200 meters, with a muzzle velocity of 650 m/sec.

Antitank shell (C/1916), weight 427.3 kg, length 1325 mm. With a propellant charge of 153 kg of BM 16 powder, the antitank shell, with a muzzle velocity of 867 m.sec, reached a maximum range of 33,200 meters.